For my husband Tondi and children. — M.G.H.

To the Galloway, Taylor, Jennings, Hunter and Aiku families, for your inspirational support. — M.G.H.

Library of Congress Control Number: 2012919378

ISBN-13: 978-1480074149
ISBN-10: 1480074144

Printed in the U.S.A.

Art direction by Diane Taylor and Bruce Hunter

Morning Star

Dr. Michelle Galloway-Hamani

Author's Letter

I did not recognize my zeal for Chemistry until the later years in high school. The object that captured the apple of my eye was the splendor of colors following chemical reactions. Consequently, I acquired a Ph.D. and fulfilling career in Chemistry. Largely, I learned sowing fruitful seeds in the younger generations brought instant gratification. To channel this gratification, writing books became the window-opportunity. Therefore, the aim of *Morning Star* is to introduce General Chemistry and Physics as it relates to understanding key characteristics of the sun and its scientific importance.

Rise and shine, Morning Star.

Hello Sun!

A bright ball of gas (g) full of hydrogen (H) and helium (He) atoms.

These H atoms join to make a He atom while releasing a **bright light** and energy in the form of heat (Δ).

Common Reaction in Stars

{ H + H + H + H $\rightarrow$ He + **bright light** + Δ }

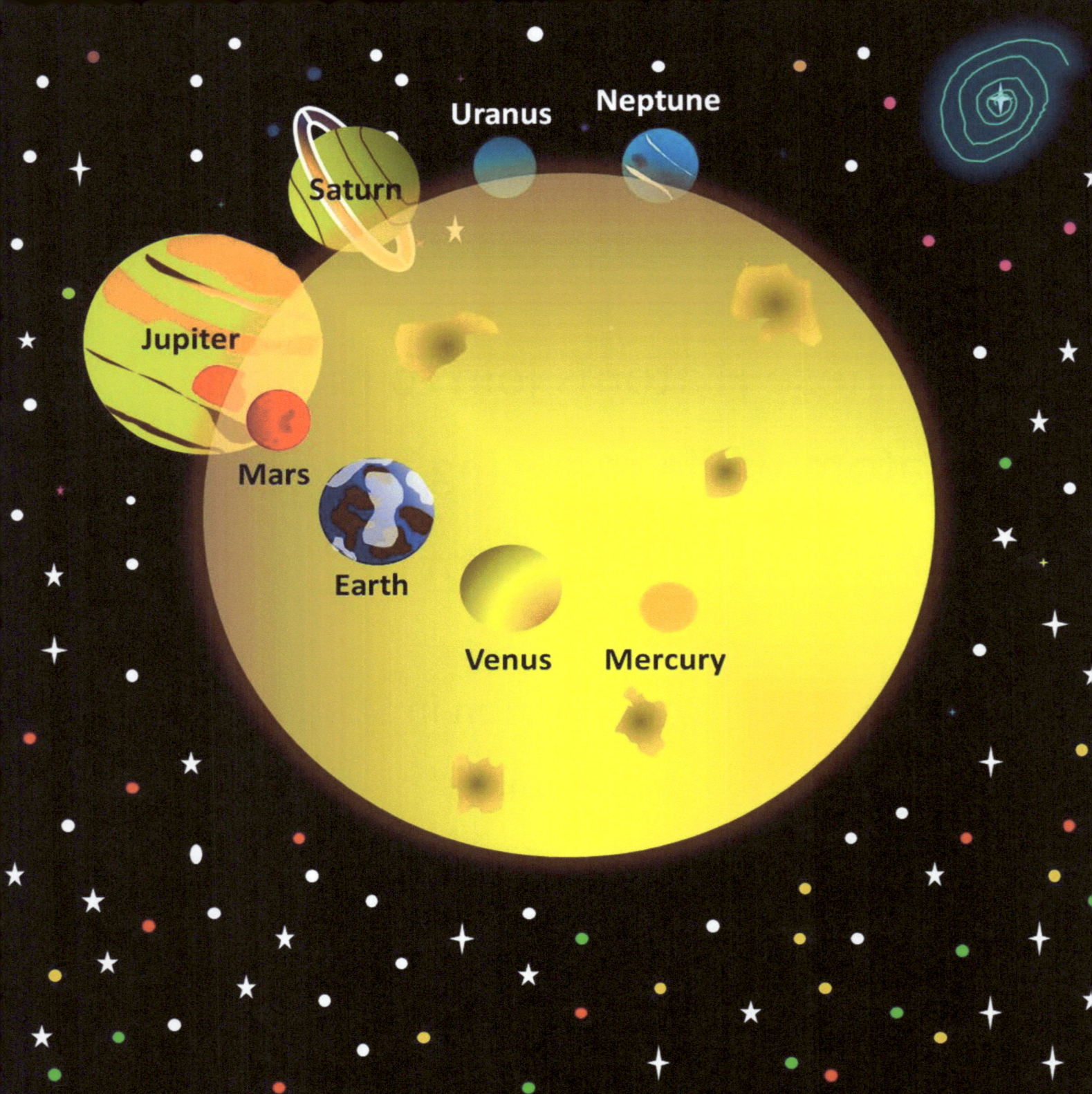

The solar system energy center of which planets orbit.

Goodnight, Morning Star.

Hello Moon!

Science Nation Questions

1. What is the title of the story?

2. From your reading, is the sun or moon a star?

3. Match each word to the correct science symbol.

gas	Δ
heat	He
hydrogen	(g)
helium	H

4. What is an atom? Construct your idea of the hydrogen and helium atoms using the activity cut out cards.

5. Select the items below that contain similar gaseous atoms as the sun.

balloon	neon lights	air
soap	rocket fuel	soda

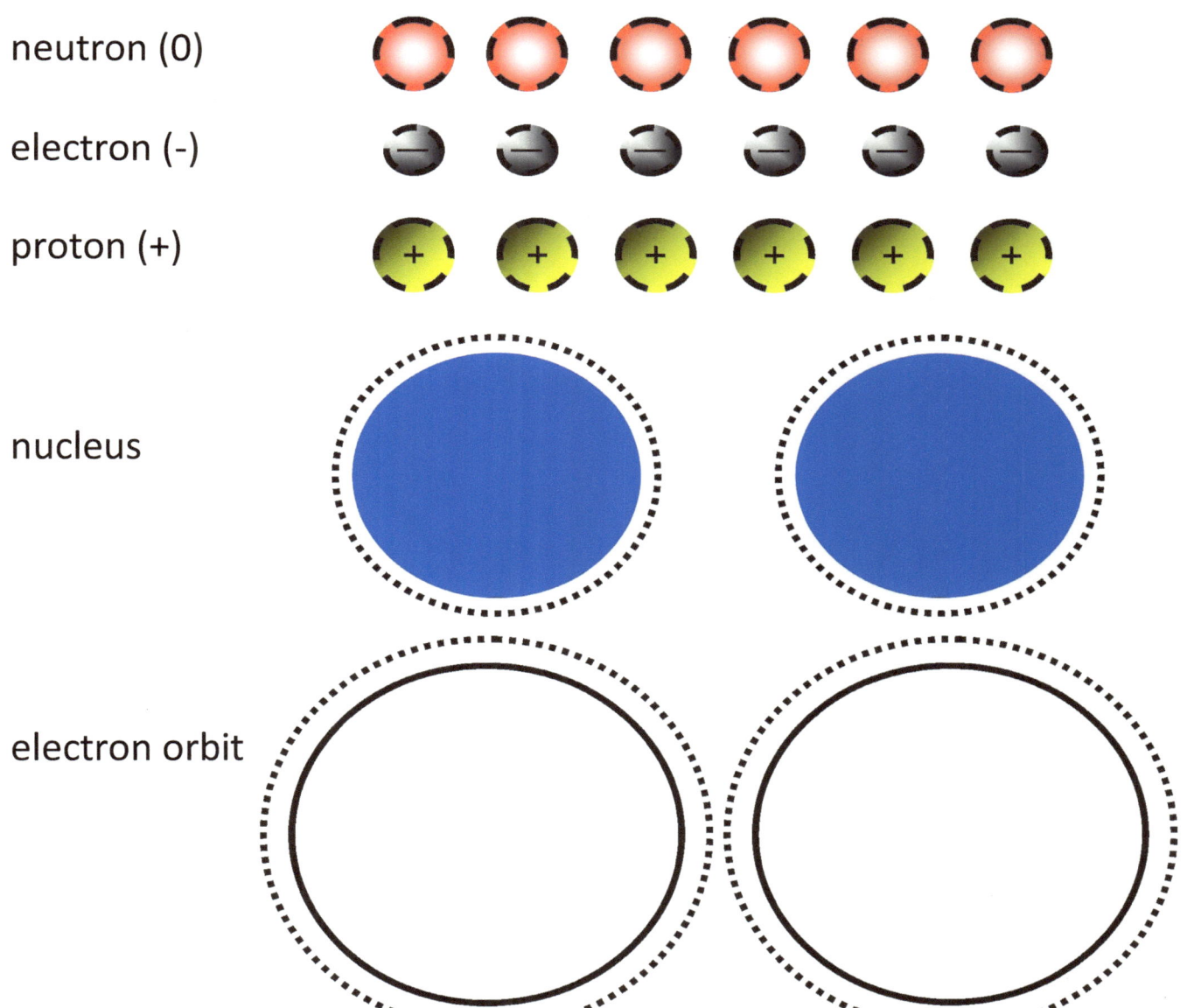

neutron (0)

electron (-)

proton (+)

nucleus

electron orbit

Science Nation Answers

1. Morning Star

2. The sun

3. gas is (g), heat is Δ, hydrogen is H and helium is He

4. Atom is the smallest building part of matter

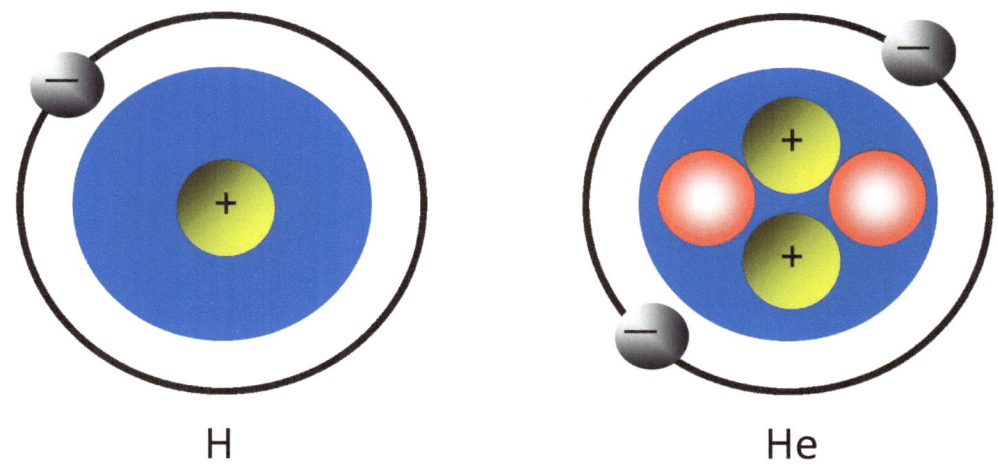

H   He

5. balloon, neon lights, air, rocket fuel and soda

Glossary

atom – smallest building part of matter

delta – (symbol Δ) energy in the form of heat created by atoms moving fast

electron – (symbol –) a negative charged particle

electron orbit – a path that electrons travel

gas – (symbol g) a state of matter with no fixed shape or volume

helium – (symbol He) a gas atom at room temperature (68 – 77°F)

hydrogen – (symbol H) a gas atom at room temperature (68 – 77°F)

neutron – (symbol 0) a particle with no charge

Glossary

nucleus – an atom center containing protons and neutrons

orbit – a path

proton – (symbol +) a positive charged particle

star – a ball of gas

sun – a large star of hydrogen and helium atoms with a sprinkling
of oxygen, carbon, iron and neon atoms in gas forms

sun spot – a dark, cool area on the sun's surface

yields – (forward arrow symbol →) pointing from starting atoms
(left) to produce the matter formed (right) in a chemical
reaction